SMART SCIENCE

Energy

Robert Snedden

Heinemann Library
Des Plaines, Illinois

Text designed by Visual Image
Cover designed by M2
Illustrations by Paul Bale, Peter Bull and Jane Watkins
Printed in Hong Kong

03 02 01 00
10 9 8 7 6 5 4 3 2

Library of Congress Cataloging-in-Publication Data

Snedden, Robert.
    Energy / Robert Snedden.
        p. cm. -- (Smart science)
    Includes bibliographical references and index.
    Summary: Discusses various aspects of energy, including power, kinetic energy, potential energy, photosynthesis, food chains, water power, wind power, chemical energy, atomic energy, heat, and solar energy.
        ISBN 1-57572-879-6 (lib. bdg.)
    1. Power resources—Juvenile literature. 2. Force and energy--Juvenile literature. [1. Power resources. 2. Force and energy.] I. Title. II. Series.
    TJ163.23.S673 1999
    333.79—dc21   98-49860
                                                                  CIP
                                                                  AC

Acknowledgments
The publisher would like to thank the following for permission to reproduce photographs:
Action-Plus/Peter Spurrier, p. 7; Franck Faugere, p. 8; Tony Henshaw, pp. 9, 15; J. Allan Cash, p.14; Empics/Mike Egerton, p. 4; FLPA/R. Wilmshurst, p. 10; M. Clark, p. 11; F. Polking, p. 12; Mark Newman, p. 13; J. C. Allen, p. 18; Silverstris, p. 23; Science Photo Library/Alex Bartel, p. 21; Alfred Pasieka, p. 22; Jerry Mason, p. 24; John Mead, p. 25; Simon Fraser, p. 27; NASA, p. 28; Space Telescope Science Institute/NASA, p. 29; Still Pictures/Paul Gipe, p.16; Dominique Halleux, p. 17; Jorgen Schytte, p. 26; The Stock Market, p. 5; Tony Stone Images/John Millar, p. 6.

Cover photograph reproduced with permission of Science Photo Library (Alfred Pasieka)
Every effort has been made to contact copyright holders of any material reproduced in this book. Any omissions will be rectified in subsequent printings if notice is given to the Publisher.

Note to the Reader
Some words in this book are shown in bold, **like this.** You can find out what they mean by looking in the glossary.

CONTENTS

MAKING THINGS HAPPEN

Energy makes things happen. Think of all the things you can do when you feel full of energy. You can run, jump, or play football. You may even help around the house or do homework!

Energy gives power to you and to everything around you. The light and heat from the sun are forms of energy. A ball flies from your bat because you have given it energy. Electrical energy powers the television, washing machine, and all sorts of other equipment in your home and at school.

Work

Energy is the ability to do **work**. To a scientist, the word "work" means transferring energy from one place to another. When you kick a ball, you are doing work! Some of the energy stored in your muscles is transferred to the ball. It becomes movement energy, which is called **kinetic energy**. More available energy means the more work can be done. Work is measured in **joules**. One joule is roughly the work needed to lift an apple just over three feet (one meter) off the ground.

When you hit a ball with a racket, you give the ball the energy it needs to move.

Power

Power measures how fast energy is transferred from one place to another. It is the rate at which work is done. Power is measured in **watts**.

One watt is equal to one joule per second. This means that every second, a 100-watt light bulb is using enough energy to send an apple 110 yards (100 meters) into the air!

To launch a space shuttle like this one takes the same amount of power as 140 jumbo jet airplanes!

It's a Fact—Energy and You

You use around 1,000 joules of energy when you climb a flight of stairs—enough to light a bulb for 10 seconds!

Try This—Bottle Motor

You need: an empty dish soap bottle, two wooden dowels—one two in. (five cm) long and the other six in. (fifteen cm) long, a rubber band (about half the length of the bottle), and a hook made from a wire coat hanger

What to do: Remove the nozzle from the bottle and make a hole in the base. Ask an adult to help you. Push one end of the rubber band through the hole in the bottom of the bottle and hold it in place with the small dowel. Use the hanger hook to catch the other end of the rubber band. Pull it through the neck of the bottle. Hold it in place with the larger dowel. Wind the dowel a few times and then let your bottle roll!

EVER-CHANGING ENERGY

Energy is always changing from one form to another. Light energy from the sun is stored as **chemical energy** in plants. You eat some plants as food, and their chemical energy becomes heat to keep your body warm.

In a busy kitchen, energy is used to prepare food, which will provide chemical energy when eaten.

Energy Conservation

No matter how it changes, the total amount of energy in the universe is always the same. Energy cannot be created or destroyed. It can only be changed.

Sometimes energy seems to disappear or get used up. No matter how hard you bounce a ball, it will stop bouncing eventually. Where does the energy go? When you start bouncing the ball, you give it **kinetic energy**. Some of the kinetic energy is converted into heat. The ball warms the air very slightly as it pushes through the air. The ground and the ball itself are warmed by the impact of the ball. Some energy is converted into the sound you hear when the ball hits the ground. Eventually, all of the kinetic energy of the ball has been converted into other forms, and it stops moving.

It All Adds Up

If you could measure the different amounts of energy that went into heating the air, the ground, and the ball, and into producing the sounds you heard, it would come to the same amount of energy you originally used to bounce the ball. Energy is never lost, although it may take a form that is difficult for us to make use of.

These rowers are converting the chemical energy in their muscles into the kinetic energy that sends them racing through the water.

It's a Fact—A Joule in his Crown!

In the 1830s, James Joule began a series of experiments. He measured how much heat different activities produced. Although not a trained scientist, Joule became the first to explain that energy cannot be created or destroyed, but only changes its form. We now measure energy in units called **joules**, which are named after him.

Try This—Spinning Snakes

You need: circles of colored paper, scissors, thread, and a radiator or other heat source

What to do: Carefully cut the circles into spiral shapes. Draw a snake's head at one of the spiral's ends. Attach some thread to the center of the spirals and hang the snakes above the radiator. Watch them twist and turn in the warm air. Heat from the radiator becomes movement energy in the snakes.

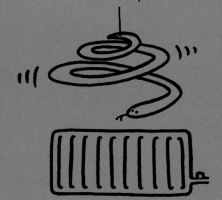

POTENTIAL AND KINETIC ENERGY

Have you ever coasted downhill on a bicycle? As you went whizzing down the hill, you were changing **potential energy** into **kinetic energy**, the energy of movement.

Earned Energy

When you lift up an apple or ride a bike to the top of a hill, you are doing **work**. You are changing the **chemical energy** stored in your muscles into kinetic energy. While you are doing this, you are storing another form of energy—potential energy.

Potential energy is energy that may be used later to do work. Whenever you lift an object, whether it is raising an apple to your mouth or riding your bike up a hill, you are working against the force of **gravity**. Given the chance, gravity will pull the object back down again. The potential energy stored in the objects becomes kinetic energy as the apple falls or the bike speeds down the hill.

A skier changes potential energy into kinetic energy as he speeds downhill.

When you wind a clock, you are storing potential energy in the spring. As the spring uncoils, the potential energy is gradually changed into kinetic energy, as it moves the hands of the clock.

Mechanical Energy

Because they are so closely related to each other, potential energy and kinetic energy are sometimes considered together as mechanical energy.

Potential energy in the taut bowstring is suddenly turned into kinetic energy when the arrow is released.

It's a Fact—Speed and Energy

Quickly-moving objects have a lot of energy. Spacecraft returning to Earth have a large amount of kinetic energy because they travel so fast. This is changed into heat energy as the spacecraft travels back through the **atmosphere.** Heatshields protect the crew from the high temperatures.

Try This—Rolling, Rolling, Rolling

You need: a toy car, a pile of books, and a piece of wood or stiff cardboard big enough to roll the car along

What to do: Prop the wood against a stack of books to make a slope and roll the car down. How far does it go? Try rolling the car down slopes of different heights. Does it make a difference how steep the slope is? Or is it the height that is important? You should find that the higher you place the car, the more potential energy it has.

9

CAPTURING THE SUN'S ENERGY

Even if you aren't a vegetarian, all the food you eat can be traced back to green plants. These green plants captured the energy of the sun and turned it into **chemical energy**.

It's Good to be Green!

All green plants, from the tiniest **algae** to the tallest trees, make their own food. In a process called **photosynthesis**, plants use carbon dioxide gas, water, and the sun's light energy to make **glucose**. **Glucose is** a type of sugar. The plant uses this sugar to live and grow. It stores the sugar in the form of **carbohydrates** until it is needed, or uses the sugar to provide a store of energy for its seeds. Along with many other animals, we eat these stored sugars—which are a source of energy—as fruits, nuts, grains, or vegetables.

Plants turn toward the sun to get the maximum amount of light on their leaves.

Why Plants are Green

Plants contain a remarkable substance called **chlorophyll**. Most of it is stored in their leaves. Chlorophyll absorbs red and blue light and reflects green light. That's why plants are green. It is chlorophyll that captures the energy of sunlight that the plant uses to make sugar.

Pass It On!

Green plants are very important in nature. They are the only living things, apart from some bacteria, that make their own food. The energy plants capture from the sun travels through the living world. The energy passes from one living thing to another as animals eat plants and are then eaten by other animals.

When we pick and eat fruit, we are helping ourselves to the sun's energy that has been stored by the plant.

It's a Fact—Photosynthetic Abundance

Cultivated land throughout the world produces around nine billion tons of plant material every year for humans and animals to eat. This is just five percent of the total amount of plants that grow every year over the whole earth. All of this staggering growth is due to photosynthesis.

Try This—Life from Light

You need: a house plant, a cardboard box (big enough to cover the plant) with a narrow vertical slit cut in one side and a sunny window

What to do: Water the plant well. Then place the box over it so light can only come in through the slit. Put the box and plant near a window and leave it for a few days. Then look at the plant. You should see that it has grown toward the light to get the energy it needs.

FOOD CHAINS

Plants are called **producers** because they produce the energy for practically all other living things. They are the first link in a **food chain**. Energy flows through the food chain, beginning with the sun, to the plants, to the animals that eat the plants and each other.

The energy from plants might help this antelope escape—or it might become energy for the cheetah if the cheetah succeeds in catching the antelope.

Links in the Chain

When a caterpillar eats a leaf, it uses some of the energy stored in the leaf as fuel to live, move, and grow. Some of the energy captured from the sun by the plant is now stored in the body of the caterpillar. The caterpillar is called a **consumer**. It consumes, or makes use of, the plant material. If a bird eats the caterpillar, the bird will be able to make use of the energy stored in the caterpillar. Again, some energy will be used up by the bird itself as it sings, flies, and hunts for more caterpillars, but some will be stored in its body. If the bird is caught by a cat, the cat in turn will get energy from eating the bird.

It's a Fact—Short Chains

Only about a tenth of the energy a plant captures from the sun will find its way into the next step of the food chain. Food chains are rarely more than five links long because of the energy loss that happens along the way.

Energy Pyramids

At every step along the chain, from plant to caterpillar to bird to cat, less and less of the energy originally captured from the sun becomes available to the animal at the next step. Animals do not just store the energy in their food; they need it to live. Animals are constantly converting energy as they move around, breathe, and so on. That energy cannot then be used by other animals. Because of this, the further along the chain you go from the plants, the fewer animals there will be. There are more birds than cats, more caterpillars than birds, and more plants than caterpillars.

Rain forest plants provide the energy to support a large number of animal species.

Try This—A Food-chain Mobile

You need: cardboard, pencils, scissors, thread, sticky tape, and a wire coat hanger

What to do: Draw and cut out eight leaves, four caterpillars, two birds, and one cat. Using tape and thread, attach the leaves to the hanger. Now attach the caterpillars to the leaves (one caterpillar to two leaves), then the birds to the caterpillars, and finally the cat to the birds.

MUSCLE POWER

Before other sources of **power** were discovered, the only energy available to do **work** came from the muscles of people and animals.

Today, the muscle power of animals is still used throughout the world.

Respiration

Food cannot be used directly to provide energy for your body. Your body cannot burn food to obtain energy as you would burn a log on a fire. Energy is released from food by combining it with oxygen in a process called **respiration**. Oxygen from the air enters your bloodstream through the lungs when you breathe in. The blood carries the oxygen to the body's cells, where it is combined with sugar, the body's fuel. This is similar to a very controlled form of burning. It releases energy that the body can use. Some of this energy is in the form of heat to keep your body at a comfortable temperature. The rest might be used to send messages along your nerves to various parts of your body, to digest more food, and to build and repair your body.

Your muscles need energy in order to work, and they, too, get their supply of energy from the food you eat.

Muscle cells contain chemicals called **enzymes** that help to break down some of the food and release its energy. Some of this energy can be stored in a special chemical in the muscle until it is needed.

Metabolism

Metabolism is the name for all of the chemical processes that go on in your body to keep it healthy and working properly. Metabolism goes on at a faster rate just after you have eaten or when you are exercising. Your metabolic rate, the rate at which you burn your food to release energy, is an important indicator of your overall health.

A trained weight lifter can deliver a tremendous amount of power from his muscles.

It's a Fact—The Going Rate!

A human's metabolic rate is about three times faster than an elephant's, but the metabolic rate of a mouse is about eight times faster than a human's.

Try This—Warming Up

You need: yourself and a press-on thermometer

What to do: Take your temperature after you have been sitting quietly for awhile. Now go for a run, dance around, or do another lively activity for a few minutes. Stop and take your temperature again. Has it gone up? You have just converted **chemical energy** in your muscles into heat energy.

WIND AND WATER

The sun provides the energy that powers the world's weather systems. It makes the wind blow. It causes water to **evaporate** from the oceans and to fall elsewhere as rain. The pull of the sun and moon on the earth's oceans produces the rising and falling of the tides. The energy of wind and water can be harnessed for our use.

Wind farms, such as this one in California, can provide a lot of energy, but they also take up a lot of space.

Blowing in the Wind

Winds blow because the sun heats some parts of the earth more than others. More of the sun's energy reaches the surface of the earth around the **equator** than at the poles. This means that it is hot at the equator and cool at the poles. The warm air at the equator expands and rises, and the cooler, denser air at the poles sinks. So the energy of the sun heating the **atmosphere** produces currents of air called wind.

Wind Power

The **power** of the wind can be captured and its energy used to do **work**. Windmills have been used for over 1,000 years to do work such as pumping water or grinding grain. Windmills have two or more blades or sails mounted onto a shaft. The sails are spun by the wind and turn the shaft to provide power. In the late19th century, windmills were in use in the United States and Europe. They were gradually replaced by other power sources, such as coal-powered steam engines and electric motors.

The world's first tidal power station, at La Rance, Brittany, France, can produce enough power to provide for the needs of around 300,000 people.

In recent times, wind farms of more than 100 wind-powered **turbines** have been set up to generate millions of **watts** of pollution-free electric power.

Water Power

Water from rivers, lakes, and seas is constantly evaporating. It becomes water vapor, a gas in the atmosphere. When the air gets colder, it becomes water again and falls as rain. In **hydroelectric** power stations, which provide a fifth of the world's electricity, rain water is collected in reservoirs. The water is then sent down pipes. The falling water provides the energy to turn the turbines that drive the **generators** that produce the electricity. Tidal power stations use the rise and fall of the tides to provide the power to drive the turbines.

It's a fact—Wind for Life

If the wind did not blow to take heat from the equator to the poles, and bring cold air from the poles, most of Earth's surface would be unsuitable for life. It would be either too hot or too cold.

Try This—Make a Windmill

You need: a square of stiff paper, scissors, sticky tape, a stick, a thumbtack, and a bead

What to do: From each corner of the square, carefully cut in about two-thirds of the way toward the center. Fold one corner of each of the triangles you have made into the center to make the windmill sails. Tape the corners together. Push the thumbtack through the corners. Then push it through the bead and into the stick.

CHEMICAL ENERGY

Fuels are substances that release a lot of energy when they are burned. This energy is stored in the chemicals that make up the fuel. The energy of almost all of the fuels we use can be traced back to the sun.

The land here is actually yielding two types of oil. Petroleum is extracted by the donkey pump, and soy oil will come from the soy beans growing in the field behind it.

Ancient Sunlight

Coal, oil, and natural gas are all **fossil fuels**. Their energy comes from the sun. These fuels were formed from the remains of plants and animals that lived many millions of years ago. Just as plants do today, the plants that lived on the earth long ago stored energy from the sun. After the plants died, they were slowly buried deeper and deeper under the ground. Over many millions of years, the incredible heat and pressure deep within the earth turned the plant remains into coal and petroleum. When we burn these fuels today, we are releasing the energy of the sun that was captured more than 300 million years ago.

It's a Fact—Greenhouse Gas

Around 5 billion tons of carbon dioxide gas are released into the **atmosphere** every year from the burning of fossil fuels. Carbon dioxide is a **greenhouse gas**. Some scientists fear that it is contributing to rising global temperatures.

Breaking the Bonds

Fossil fuels are mostly made up of the **elements** carbon and hydrogen. The carbon and hydrogen are linked together by strong bonds. A lot of energy, called **chemical energy**, is stored in these bonds. This energy can be released if the bonds are broken. Burning a fuel breaks the bonds between the chemicals that make it up, releasing its energy. The carbon and hydrogen combine with oxygen in the air to form water and carbon dioxide gas. Many people today are concerned that the huge amounts of carbon dioxide gas being released into the atmosphere by the burning of fossil fuels are changing the world's climates. Other harmful chemicals are also produced by burning fossil fuels.

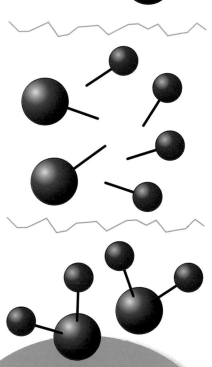
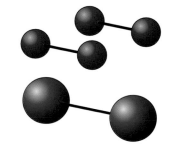

Energy is released as bonds between atoms are broken and re-formed.

Try This—Bubble Mixture

You need: vinegar, baking soda, a spoon, and a jar

What to do: Pour some vinegar into the jar. Then add a spoonful of baking soda. A chemical reaction will take place between the two, releasing bubbles of carbon dioxide gas. Bonds are broken and re-formed as the vinegar and baking soda react together. The jar also becomes a little warmer as heat energy is released.

NUCLEAR ENERGY

Nuclear energy comes from the innermost part of an **atom**—its nucleus. It is the most powerful source of energy we have.

Nuclear Power

At the heart of an atom is its nucleus. The nuclei of all **elements,** apart from hydrogen, are made of **neutrons** and **protons.** The nuclei of some large atoms are unstable. The nuclei split to form new, smaller atoms. When this happens, energy is released. Materials with unstable nuclei are called **radioactive.**

Atom

Neutron

Incoming
Neutron

Uranium—a radioactive element—is often used as a nuclear fuel. It is a metal that is mined in the United States, Australia, and in other places. When a uranium atom splits, it releases two or three neutrons, as well as energy. The neutrons collide with other uranium atoms, causing them to split. This is called a **chain reaction.** The energy from the atoms can be used to heat water to make steam to drive a **turbine** and generate electricity. Just over two pounds (one kilogram) of nuclear fuel can provide as much energy as over 6.6 million pounds (3 million kilograms) of coal.

In a nuclear chain reaction, particles from the break up of an atom trigger the break up of still more particles.

Radiation Dangers

The advantage of nuclear fuel over **fossil fuels** is that it does not produce polluting gases. However, radioactive materials are very harmful to living things. High-speed particles and energy waves given off by these materials can damage living cells and may cause cancers. Radioactive waste has to be disposed of very carefully, and strict safety rules at nuclear power stations keep workers from coming into contact with radioactive materials.

Nuclear energy raises two main concerns: the safety of the power station and the disposal of waste.

It's a Fact—A Lot of Power

Two pounds (one kilogram) of uranium could release enough energy to light a million light bulbs for over 100 days!

Try This—Domino Chain Reaction

You need: dominoes and a flat surface

What to do: Stand the dominoes on end, arranged to form a triangle. One domino should be at the point of the triangle, followed by two dominoes just behind it, then four dominoes, then eight, and then sixteen, depending on how many dominoes you have. Gently push over the first domino so that it knocks over the next two, which knock over the next four, and so on. The reaction continues until you run out of domino "fuel."

HEAT

We all need to keep warm. Heat is an important form of energy for living things. Living things need to stay within a range of **temperatures**. If they get too hot or too cold, they die.

What Is Heat?

Heat is the energy of motion of the tiny **particles** that make up matter. **Atoms**, which make up all the matter we can see, are never still. They vibrate constantly, even in objects that may appear to be perfectly still. The temperature of the object tells us how much the atoms that make it up are moving. The faster the atoms are moving, the higher the temperature will be. The more atoms there are moving, the more heat energy there is. A large object has more heat energy than a small one at the same temperature.

This thermal image shows the different temperatures of parts of the body. The hottest parts are red and the coolest are blue.

From Hot to Cold

Heat energy always moves from warm objects to cooler ones. You can usually tell whether an object's temperature is higher or lower than yours by touching it. If you touch a cold object, the heat flows from your hand into the object, making your hand feel cold. If you touch a warm object, the opposite happens. Heat flows into your hand, making it feel warm.

Heat can move from one place to another in three different ways—by **conduction**, by **convection,** or by **radiation**. Conduction happens when heat flows from one part of an object to another part, or from one object to another one that is touching it. Metals are the best **conductors** of heat. If you leave a metal spoon in hot liquid, it becomes warm as heat is conducted from the liquid to the spoon. Heat that is carried by gases and liquids moves by convection currents. The liquid or gas nearest the heat source warms up and expands, pushing and warming the liquid or gas next to it. Cooler material flows in to take the place of the expanding material and is then heated in turn. The wind is a convection current in the air caused by the heat energy from the sun. Radiation is when heat travels in waves of heat energy from hot objects. The heat from the sun reaches the earth by radiation.

The workers in this steel mill have to wear protective clothes to withstand the fierce heat of the molten metal.

It's a Fact— Hot Stuff!

Inside the sun, the temperature is about 28.8 million °F (16 million °C). A wood fire has a temperature of only 500°F (260°C).

Try This—Hot or Cold?

You need: three large bowls: one containing iced water, one with hand-hot water, and one with warm water

What to do: Put one hand in the iced water and the other in the hand-hot water. Keep them there for a minute. Now put both hands in the warm water. Does it feel hot or cold to each hand?

SOLAR ENERGY

The sun is our most important source of energy. All of the energy we use, apart from **nuclear energy** and energy from hot rocks inside the earth, can be traced back to the sun.

The sun's energy is turned into food by green plants, which provide the food for all of the earth's animals. Energy captured by plants millions of years ago is trapped in the **fossil fuels** we depend on for energy—coal, oil, and gas. The energy of the sun heating the air makes the wind blow.

Disappearing Sun

Colossal amounts of energy are given off by the sun. The sun is a huge ball of gases with a diameter 109 times that of the earth. Inside the sun, the pressure and **temperature** are so unimaginably high that the **atoms** that make up the matter in the sun are crushed together. Some of the matter of the sun is converted into energy in the process. In fact, the sun is getting lighter by around 908,000 tons every second as its matter is changed into energy. Don't worry though. The sun isn't about to vanish altogether!

Scientists at this research facility are trying to copy the energy-releasing processes that take place inside the sun.

The sun is so big that it will be about ten billion more years before it starts to run out of nuclear fuel.

Fast Heat

It takes thousands of years for the energy generated at the center of the sun to make its way up to its surface. It is carried by **radiation** and by massive **convection** currents in the hot gases that make up the sun. Once the energy reaches the sun's surface, it is radiated out into space at the speed of light—186,300 miles (300,000 kilometers) per second—and reaches the earth around eight minutes later.

These mirrored dishes can be turned to focus and gather the light of the sun to generate electricity.

It's a Fact—Sun Power

In one second, the sun gives off 13 million times more energy than the United States uses to generate all the electricity it needs for a year!

Try This—Solar Scorcher

Ask an adult to help you with this.

Also, remember that you should never look directly at the sun.

You need: a magnifying glass, a sheet of paper, a bucket half-full of sand, a jug of water, and a sunny day

What to do: Crumple the paper into the bucket. Then, with the sun behind you, hold the lens above the paper. With an adult, move the lens nearer or farther away until a bright spot of light appears. After a short time, the paper will begin to scorch and burn. Use the water in the jug to put out the fire. Can you see one reason why it is important not to leave glass lying around outside?

WILL WE RUN OUT OF ENERGY?

Most of the energy used by modern industrial nations comes from the burning of **fossil fuels**. Fossil fuels are **non-renewable** energy sources. This means that only limited amounts exist, and when these supplies run out, there will be no more.

The Search for Renewable Energy

Energy cannot be destroyed but can take forms that we can no longer make use of. Once a fuel has been burned, the heat energy is lost, and it cannot be turned back into usable energy. Some forms of energy are available in almost unlimited supply, however. These are called **renewable energy** sources.

Renewable Energy

Energy from sunlight is a renewable energy source. It is hard to imagine that we would ever use up all the energy from the sun! Solar panels can be used to collect the sun's rays for heating. Solar cells can convert light energy into electricity.

A biogas digester in India is used to convert plant and animal waste into fuel.

The power of the wind, running water, and the tides can also be harnessed to produce energy.

The deeper you go inside the earth the hotter it becomes. This **geothermal** energy is used in some parts of the world, such as Iceland, to generate electricity and to provide heat for buildings.

We can also make use of the energy stored by living things. This is called **biomass energy,** and we can obtain it from wood, which can be burned, and from animal wastes, which produce a gas we can burn when they decay. Biomass energy is not truly unlimited, however, and there are pollution problems involved with burning it.

A power station in Iceland uses geothermal energy to generate electricity. People swim in the hot pools near the power station.

It's a Fact—Water Watts
The Grand Coulee **hydroelectric** power plant in the state of Washington can generate 10,000 million **watts** of electrical power.

Try This—Be an Energy Saver!
What to do: Switch off the lights when there is no one in the room. Put on a sweater when it is cold rather than turning up the heat. Do not buy things that use a lot of packaging. Make sure that newspapers and bottles are recycled. What other ways can you find to save energy?

THE ORIGINS OF ENERGY

If energy cannot be created or destroyed, but just goes on changing, where did all the energy in the universe come from? Will it ever stop changing?

The Big Bang

A long, long time ago there was no Earth, no sun, and no stars. Hard as it is to imagine, scientists believe that everything in the universe was once packed into a space that was smaller than the smallest **particle** in an **atom**.

Suddenly, this point that contained the universe erupted and began to expand. The universe is still expanding today. All of the energy in the universe was contained in this explosion. The **temperature** of the early universe was far greater than anything that exists today. It was so hot that matter could not exist at all. It was only when the young universe began to cool that the particles that make up matter began to appear. They formed from some of the energy in the universe, rather like drops of water appearing from cooling steam.

This image, built up by the COBE satellite, shows ripples of energy in space—the remnants of the big bang.

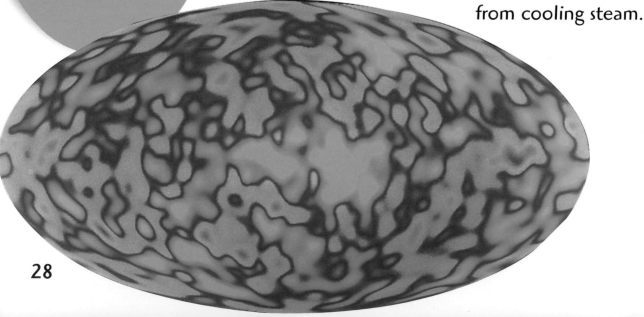

Big Crunch or Heat Death?

Scientists argue about what might happen to the universe in the far future. One possibility is that the expanding of the universe following the **big bang** explosion might be slowed down and stopped and then reversed by **gravity**. Everything will start to collapse together until, once again, all matter and energy in the universe are concentrated into a vanishingly small point.

Another possibility is that the universe will continue to expand on and on forever. Eventually, all energy will be transformed into heat, and nothing else will happen after that.

Do not worry about which of these possibilities might occur. It will not be for a very long time— several tens of billions of years, in fact!

The Hubble Space Telescope has revealed galaxy upon galaxy of stars, all speeding away from us as the universe expands.

It's a Fact—Hot Stuff!

Just after the big bang, the temperature of the universe was about 100 million million million million million °F!

Try This—Expanding Universe

You need: a round balloon and a marker pen

What to do: Blow the balloon up a little and make some marks on it to represent galaxies of stars. Continue to blow the balloon up and watch how the "galaxies" appear to be moving away from each other. The same thing is happening in the universe as space expands, carrying the galaxies with it.

GLOSSARY

algae very simple microscopic plants found in water. Algae do not have flowers, leaves, or roots.

atmosphere layer of gases that surrounds the earth; the air we breathe

atom smallest part of a substance to exist; when two or more atoms join, they make a molecule

big bang the sudden explosion of energy from an incredibly tiny single point that started the universe

biomass energy energy that comes from plant material or animal wastes, such as burning wood

carbohydrate energy-giving compound made by plants and found in foods such as rice and potatoes

chain reaction reaction that continues because the last part of one reaction starts the beginning of another reaction

chemical energy energy that is stored in the bonds that hold molecules together

chlorophyll compound found in plants that absorbs light and uses its energy to make sugars in **photosynthesis.** Chlorophyll gives plants their green color.

conduction movement of electricity or heat through a substance

conductor substance through which heat passes

consumer living thing that cannot make its own food but has to eat other living things

convection movement of heat through a liquid or gas by currents

electrical energy energy that comes from a flow of charged particles

element substance that is made up of just one type of atom; it cannot be broken down by chemical reactions

enzyme chemical found in living things that can change the speed of natural chemical processes

equator imaginary line around the earth, dividing the northern and southern hemispheres

evaporate change into a vapor or gas

food chain path that energy follows from plants, to animals, to animals. Also, a series of living things, each of which eats the one that comes before it in the chain.

fossil fuels fuels—oil, coal, and natural gas—formed over millions of years from the remains of ancient plants and animals

generator machine for converting **kinetic energy** into electricity

geothermal energy   energy from the hot rocks inside the earth

glucose   simplest form of sugar, made by plants in photosynthesis

gravity   force of attraction between all objects. The bigger the object, the bigger the attractive force it produces. The earth's gravity is what pulls you back to the ground when you jump up into the air.

greenhouse gases   gases that trap heat reflected from the earth's surface, such as carbon dioxide

hydroelectric/hydroelectricity electricity produced by using the kinetic energy of flowing water

joule   unit of energy

kinetic energy   energy of a moving object

metabolism   variety of chemical processes that go on in a living thing

neutron   one of the particles that make up the nucleus of an atom

non-renewable energy   source of energy, such as fossil fuels, that can't be replaced once it has been used up

nuclear energy   energy from the nucleus (central part) of an atom

particle   tiny pieces of matter

photosynthesis   process by which plants make glucose from carbon dioxide and water, using the energy of the sun

potential energy   stored energy that an object has because of its position— for example, something held at a height or a coiled spring has potential energy

power   rate at which energy is transferred from one place to another, usually measured in watts

producer   living things, such as green plants, at the start of the food chain, that supply the energy for all other living things in the chain

proton   one of the particles that make up the nucleus of an atom

radiation   streams of high energy waves and **particles** given off by **radioactive** materials

radioactive   giving off radiation

renewable energy   energy source that will not run out, such as solar energy, the energy of the sun

respiration   process of taking in oxygen to be used to release the energy from food

turbine   machine that spins, usually using a jet of steam, and is used to drive a generator

watt   unit of power (1 joule per second)

work   transferring energy from one place to another to move an object or to change it in some way

More Books to Read

Bailey, Donna. *Energy from Oil & Gas.* Chatham, NJ: Raintree Steck-Vaughn. 1990.

Blackbirch Graphics Staff. *Alternate Energy Sources.* New York: Henry Holt & Company. 1996.

Blackbirch Graphics Staff. *Recycling.* New York: Henry Holt & Company, 1996.

Dineen, Jacqueline. *Natural Energy.* Chatham, NJ: Raintree Steck-Vaughn Publishers. 1994.

Dineen, Jacqueline. *Oil, Gas, & Coal.* Chatham, NJ: Raintree Steck-Vaughn. 1995.

Fowler, Allan. *Energy from the Sun.* Danbury, CT: Children's Press. 1997.

Gutnik, Martin J. & Natalie Brown-Gutnik. *The Energy Question: Thinking about Tomorrow.* Springfield, NJ: Enslow Publishers. 1993.

Lafferty, Peter. *Heat & Cold.* Tarrytown, NY: Marshall Cavendish. 1995.

Oxlade, Chris. *Energy & Movement.* Danbury, CT: Children's Press. 1999.

Riley, Peter D. *Food.* Des Plaines, IL: Heinemann Library. 1998.

Rising, Trudy & Peter Williams. *Light Magic: And Other Science Activities about Energy.* Buffalo, NY: Firefly Books, Limited. 1994.

Wood, Robert W. *Heat Fundamentals.* Broomal, PA: Chelsea House. 1997.

Woodruff, John. *Energy.* Chatham, NJ: Raintree Steck-Vaughn. 1998.

Index